Table of Contents

PREFACE

On March 28, 2014 the Obama Administration released a key element called for in the President's Climate Action Plan: a Strategy to Reduce Methane Emissions. The strategy summarizes the sources of methane emissions, commits to new steps to cut emissions of this potent greenhouse gas, and outlines the Administration's efforts to improve the measurement of these emissions. The strategy builds on progress to date and takes steps to further cut methane emissions from several sectors, including the oil and natural gas sector.

This technical white paper is one of those steps. The paper, along with four others, focuses on potentially significant sources of methane and volatile organic compounds (VOCs) in the oil and gas sector, covering emissions and mitigation techniques for both pollutants. The Agency is seeking input from independent experts, along with data and technical information from the public. The EPA will use these technical documents to solidify our understanding of these potentially significant sources, which will allow us to fully evaluate the range of options for cost-effectively cutting VOC and methane waste and emissions.

1.0   INTRODUCTION

The oil and natural gas exploration and production industry in the U.S. is highly dynamic and growing rapidly. Consequently, the number of wells in service and the potential for greater air emissions from oil and natural gas sources is also growing. There were an estimated 504,000 producing gas wells in the U.S. in 2011 (U.S. EIA, 2012a), and an estimated 536,000 producing oil wells in the U.S. in 2011 (U.S. EIA, 2012b). It is anticipated that the number of gas and oil wells will continue to increase substantially in the future because of the continued and expanding use of horizontal drilling combined with hydraulic fracturing (referred to here as simply hydraulic fracturing).

Due to the growth of this sector and the potential for increased air emissions, it is important that the U.S. Environmental Protection Agency (EPA) obtain a clear and accurate understanding of emerging data on air emissions and available mitigation options. This paper presents the Agency's understanding of emissions and available control technologies from a potentially significant source of emissions in the oil and natural gas sector.

Oil and gas production from unconventional formations such as shale deposits or plays has grown rapidly over the last decade. Oil and natural gas production is projected to steadily increase over the next two decades. Specifically, natural gas development is expected to increase by 44% from 2011 through 2040 and crude oil and natural gas liquids are projected to increase by approximately 25% through 2019 (U.S. EIA, 2013). The projected growth is primarily led by the increased development of shale gas, tight gas, and coalbed methane resources utilizing new production technology and techniques such as horizontal drilling and hydraulic fracturing. According to the U.S. Energy Information Administration (EIA), over half of new oil wells drilled co-produce natural gas (U.S. EIA, 2013). Based on this increased oil and gas development and the fact that half of these new oil wells co-produce natural gas, the potential exists for increased emissions from production through distribution of natural gas from these operations.

Compressors have been identified as an emission source that has the potential to produce emissions to the atmosphere during oil and gas production (gathering and boosting), processing,

2

transmission and storage. Compressors are mechanical devices that increase the pressure of natural gas and allow the natural gas to be transported from the production site, through the supply chain, and to the consumer. Vented emissions from compressors occur from seals (wet seal compressors) or packing surrounding the mechanical compression components (reciprocating compressors) of the compressor. These emissions typically increase over time as the compressor components begin to degrade. Leak emissions from various compressor components can also occur, but those emissions are not covered in this paper because the causes and mitigation techniques are different than the vented emissions.

The purpose of this paper is to summarize the EPA's understanding of vented VOC and methane emissions from compressors, and the EPA's understanding of available mitigation techniques (practices and equipment) to reduce vented emissions from compressors. Included in the mitigation techniques discussion is our understanding of the efficacy and cost of these technologies and the prevalence of use of the technologies in the industry.

In the oil and natural gas sector, the most prevalent types of compressors used are reciprocating and centrifugal compressors. For the purposes of this paper, a reciprocating compressor is defined as:

A piece of equipment that increases the pressure of a process gas by positive displacement, employing linear movement of the driveshaft.

For the purposes of this paper, a centrifugal compressor is defined as:

Any machine for raising the pressure of a natural gas by drawing in low pressure natural gas and discharging significantly higher pressure natural gas by means of mechanical rotating vanes or impellers.

Compressors are used in all aspects of natural gas development. In the production segment, compressors are used at the wellhead to compress gas for fluids removal and pressure equalization with gathering equipment systems. However, the primary use of compressors is in the natural gas processing, transmission and storage (particularly underground storage) segments of the industry.

Section 2 of this document provides background and context for discussions of vented emissions from compressors, Section 3 presents our understanding of vented VOC and methane emissions from compressors, and Section 4 provides our understanding of available emissions mitigation techniques. Section 5 summarizes the EPA's understanding based on the information presented in Sections 3 and 4, and Section 6 presents a list of charge questions for reviewers to assist us with obtaining a more comprehensive understanding of vented VOC and methane emissions from compressors and emission mitigation techniques.

2.0   PROCESS DESCRIPTIONS

2.1   Reciprocating Compressors

In a reciprocating compressor, natural gas enters the suction manifold, and then flows into a compression cylinder where it is compressed by a piston driven in a reciprocating motion by the crankshaft powered by an internal combustion engine. For the purposes of this paper, reciprocating compressor rod packing is defined to mean:

> A series of flexible rings in machined metal cups that fit around the reciprocating compressor piston rod to create a seal limiting the amount of compressed natural gas that escapes to the atmosphere.

Over the operating life of the compressor, the rings become worn and the packing system will begin to wear resulting in higher leak rates. Emissions from packing systems originate from mainly four components; the nose gasket, between the packing cups, around the rings and between the rings and the shaft. See Figure 2-1 for a depiction of a typical compressor rod packing system configuration. Typically, gases leaked from the packing system are vented.

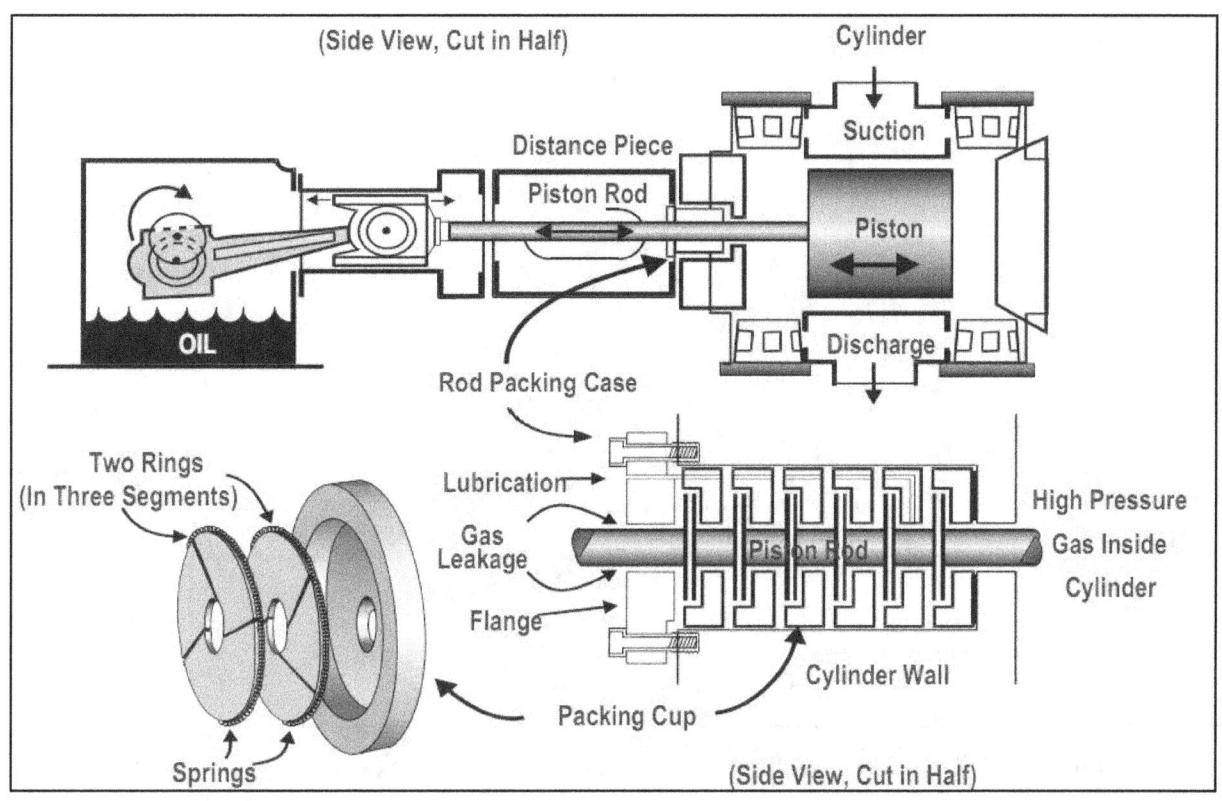

**Figure 2-1. Typical Reciprocating Compressor Rod Packing System
(U.S. EPA, 2006a)**

2.2 Centrifugal Compressors

Centrifugal compressors use a rotating disk or impeller to increase the velocity of the gas where it is directed to a divergent duct section that converts the velocity energy to pressure energy. These compressors are primarily used for continuous, stationary transport of natural gas and are widely used in the processing and transmission industry segments. Centrifugal compressors are equipped with either a wet or dry seal configuration. Wet seals use oil around the rotating shaft to prevent natural gas from escaping where the compressor shaft exits the compressor casing. The oil is circulated at high pressure to form a barrier against compressed natural gas leakage. The circulated oil entrains and absorbs some compressed natural gas that may be released to the atmosphere during the seal oil recirculation process (degassing or off-gassing). Figure 2-2 illustrates the wet seal compressor configuration.

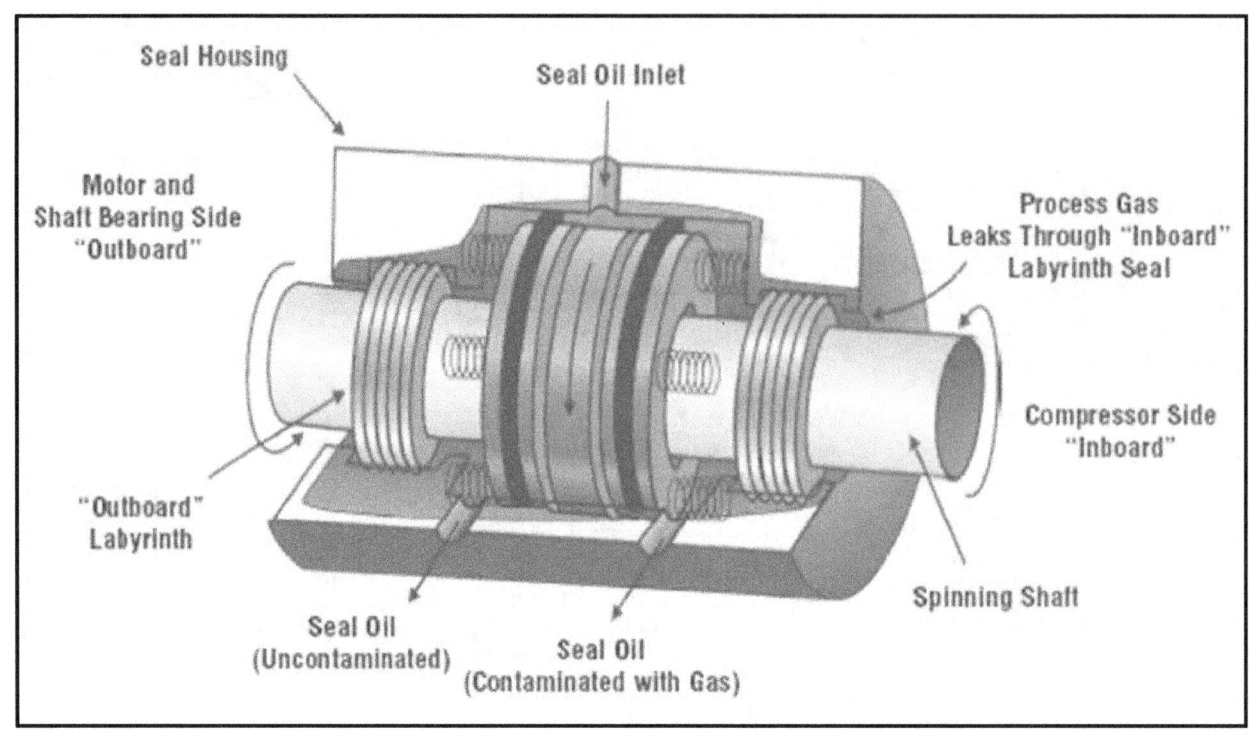

Figure 2-2. Typical Centrifugal Compressor Wet Seal (U.S. EPA, 2006b)

Alternatively, dry seal compressors use the opposing force created by hydrodynamic grooves and springs to provide a seal. The opposing forces create a thin gap of high pressure gas between the rings through which little gas can leak. The rings do not wear or need lubrication because they are not in contact with each other. The combination of two or more of the dry seals in series is called "tandem dry seals" and is effective in reducing gas leakage. Figure 2-3 illustrates the tandem dry seal compressor configuration.

Gas emissions from wet seal centrifugal compressors have been found to be higher than dry seals compressors primarily due to the off-gassing of the entrained gas from the oil. This gas is not suitable for sale and is either released to the atmosphere, flared, or routed back to a process. In addition to lower gas leakage (and therefore lower emissions), dry seals have been found to have lower operation and maintenance costs than wet seal compressors because they are a mechanically simpler design, require less power to operate, are more reliable and require less maintenance. Dry seal compressors will be discussed in more detail in Section 4.

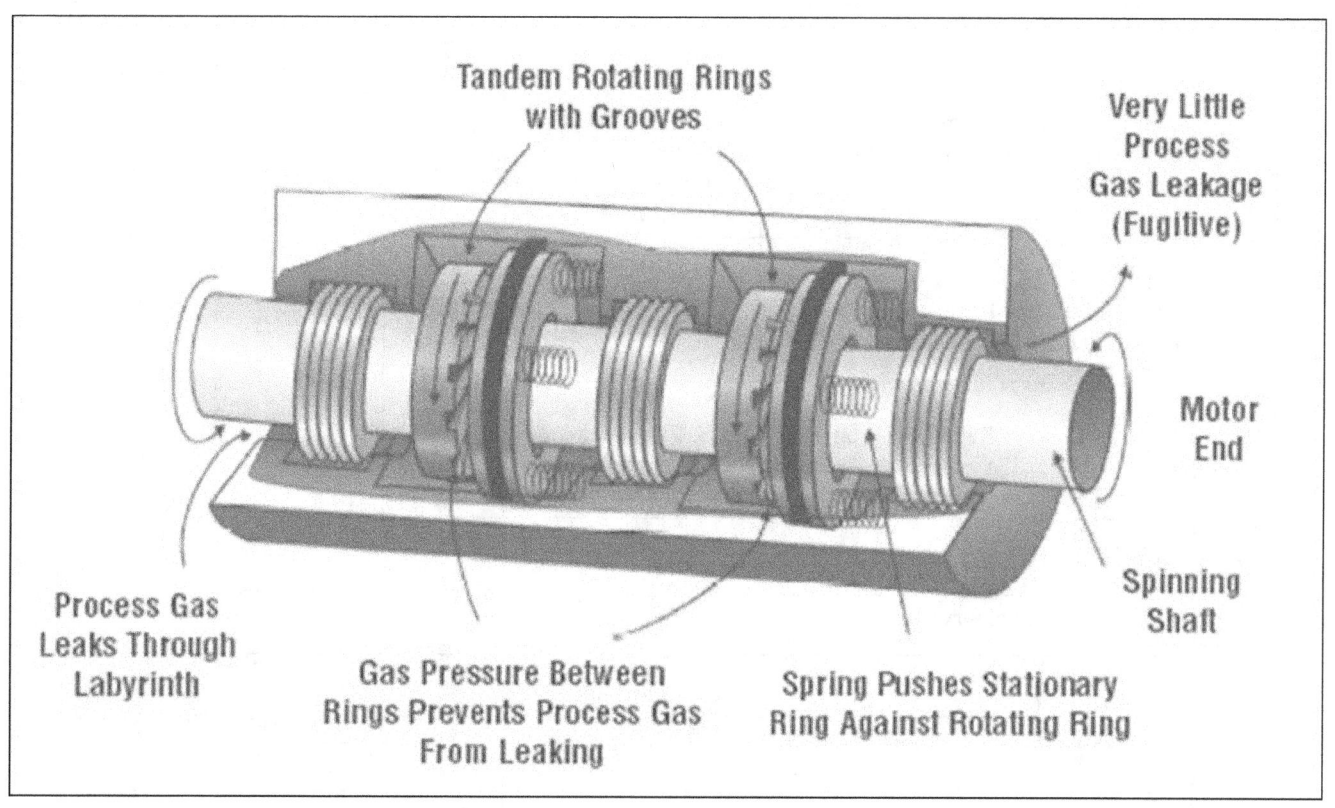

Figure 2-3. Typical Centrifugal Compressor Tandem Dry Seal (U.S. EPA, 2006b)

3.0   EMISSIONS DATA AND EMISSIONS ESTIMATES

There are several sources of emissions factors, activity data, and direct measurement data that have been used to estimate emissions from compressors in the oil and natural gas sector. Some of these studies are listed in Table 3-1, along with an indication of the type of information contained in the study (i.e., activity level and emissions data).

Table 3-1. Summary of Major Sources of Information and Data on Compressors

Name	Affiliation	Year of Report	Activity Factor	Emissions Data
Methane Emissions from the Natural Gas Industry: Equipment Leaks (GRI/U.S. EPA, 1996)	Gas Research Institute (GRI)/ U.S. Environmental Protection Industry	1996	Nationwide	X
Natural Gas Industry Methane Emission Factor Improvement Study ((URS/UT, 2011)	URS Corporation, UT Austin, and U.S. Environmental Protection Agency	2011	None	EF Only
Greenhouse Gas Reporting Program (U.S. EPA, 2013)	U.S. Environmental Protection Agency	2013	Facility Level	X
Inventory of Greenhouse Gas Emissions and Sinks: 1990-2012 (2014 GHG Inventory)	U.S. Environmental Protection Agency	2014	Nationwide	X
Analysis under subpart OOOO (U.S. EPA, 2012a)	U.S. Environmental Protection Agency	2012	Nationwide	X
Characterizing Pivotal Sources of Methane Emissions from Natural Gas Production: Summary and Analysis of API and ANGA Survey Responses (API/ANGA Survey)	American Petroleum Institute (API)/America's Natural Gas Alliance (ANGA)	2012	Regional	X[a]
Economic Analysis of Methane Emission Reduction Opportunities in the U.S. Onshore Oil and Natural Gas Industries (ICF/EDF Study)	ICF International (Prepared for the Environmental Defense Fund)	2014	Regional	X

a. The API/ANGA study provided information on equipment counts that could augment nationwide emissions calculations. No source emissions information was included.

The following sections describe emissions data, emission factors, the origin of the emission factors, and the methodologies used in the emission estimation process including the identification of national populations for several sources of information.

3.1   GRI/EPA Methane Emissions from the Natural Gas Industry, Volume 8: Equipment Leaks (GRI/EPA, 1996a)

This report provides an estimate of annual methane emissions from reciprocating and centrifugal compressor seals from the natural gas production, processing, transmission and storage sector using the component method. The component method uses average emission factors for reciprocating and centrifugal compressor seals and the average number of reciprocating and centrifugal compressors per facility to estimate the average facility emissions. The average facility emissions were then extrapolated to a national estimate using the number of facilities in each of the sectors.

The emissions data for natural gas production sites were based on screening and bagging data collected at 12 oil and gas production sites in the Western U.S. Screening involves using a handheld organic vapor analyzer (OVA) or toxic vapor analyzer (TVA) to measure the concentration (e.g., parts per million volume, ppmv) of the vented vapors. The method of bagging involves enclosing the component to collect venting vapors and measuring the flow rate. The measured flow rates from bagged equipment coupled with screening values are used to determine the unit-specific mass emission rate. A total of 40 reciprocating compressor seals were screened and bagged and an emission factor of 2.37 thousand standard cubic feet per cylinder per year (Mscf/cyl-yr) was calculated. No centrifugal compressors were located at any of the production sites that were screened.

The reciprocating and centrifugal compressor seal emissions data for the natural gas processing, natural gas transmission and natural gas storage sectors were obtained using a GRI Hi-Flow™ (trademark of the Gas Research Institute) sampler to quantify emissions and to develop emissions factors (GRI/EPA, 1996b). The sampler has a high flow rate and generates a flow field around the component that captures the entire leak. As the sample stream passes through the instrument, both the flow rate and the total hydrocarbon (THC) concentration are measured. The mass emission rate was then determined using these measurements. Different emission rates were calculated for the different operating modes of the compressor (GRI/EPA, 1996b), and were as follows;

- Operating and pressurized;
- Idle and fully pressurized;
- Idle and partially pressurized using a fuel saver system (reciprocating compressors only);
- Idle and depressurized.

The pressurized compressor seal emission rates (operating and idle) were calculated as the average of all reciprocating and centrifugal compressor seals combined. The compressor seal emission rates were determined to be 599 thousand standard cubic feet per seal per year (Mscf/seal-yr) in the operating and pressurized mode, 531 Mscf/seal-yr in the idle and pressurized mode, and 116 Mscf/seal-yr in the idle and partially pressurized mode (e.g., fuel saver) (GRI/EPA, 1996b). The compressor seal emission rate was assumed to be negligible in the idle and depressurized mode.

Using the percentage of the time pressurized and the compressor seal emission rates for the operating modes (e.g., operating and pressurized, idle and pressurized, idle and partially pressurized, idle and depressurized), emission factors were calculated for the natural gas processing, transmission, and storage segments. A summary of the emissions factors for each of these segments and the natural gas production segment are provided in Table 3-2. The number of seals was determined by averaging the compressor seal counts from the data in each of the segments. The number of centrifugal compressor seals depends on the type of compressor: centrifugal compressors with overhung rotors have one seal and beam type compressors have two seals. Information from three compressor vendors and one compressor seal vendor showed an even split between the two type of centrifugal compressors; therefore, the number of seals per centrifugal compressor was averaged to be 1.5.

**Table 3-2. Summary of Reciprocating and Centrifugal Compressor Seal
Methane Emission Factors**

Type of Compressor	Percentage of Time the Compressor is Pressurized (%)	Compressor Seal Methane Emission Factor (Mscf/seal-yr)	Assumed Number of Seals per Compressor	Average Compressor Methane Emission Factor from Seals (Mscf/yr)
Natural Gas Production				
Reciprocating	N/A	2.37	4	9.48
Natural Gas Processing				
Reciprocating	89.7	450	2.5	1,125
Centrifugal	43.6	228	1.5	342
Natural Gas Transmission				
Reciprocating	79.1	396	3.3	1,307
Centrifugal	24.2	165	1.5	248
Natural Gas Storage				
Reciprocating	67.5	300	4.5	1,350
Centrifugal	22.4	126	1.5	189

The GRI/EPA study presented the emissions for reciprocating and centrifugal compressors as a sum of the emission components from compressors. These components included methane emissions from compressor seals, blowdown open-ended line, pressure relief valves, starter open-ended line, and miscellaneous, which includes valves and connectors. For the purposes of this paper, only the methane emissions from reciprocating and centrifugal compressor seals were calculated using the equipment counts of reciprocating and centrifugal compressors and applying the methane emission factor for each of the sectors. A summary of these emissions are presented in Table 3-3 for each of the sectors reported in the GRI/EPA study.

Table 3-3. Summary of GRI/EPA Methane Emissions from Reciprocating and Centrifugal Compressor Seals

Type of Compressor	Average Methane Emission Factor (Mscf/yr)	Activity Factor, Compressor Count	Annual Methane Emissions (Mscf/yr)	Average Methane Emissions (MT/yr)
Natural Gas Production				
Reciprocating	9.48	17,152	162,601	3,071
Natural Gas Processing				
Reciprocating	1,125	4,092	4,603,500	86,949
Centrifugal	342	726	248,292	4,690
Natural Gas Transmission				
Reciprocating	1,307	6,799	8,886,293	167,841
Centrifugal	248	681	168,888	3,190
Natural Gas Storage				
Reciprocating	1,350	1,396	1,884,600	35,596
Centrifugal	189	136	25,704	485
Total			15,978,655	301,799

The GRI/EPA study reported methane emissions of 568,670 Mscf/yr (10,741 MT) from reciprocating compressors from both Eastern and Western U.S. natural gas production. These totals, as stated earlier, include emissions from compressor blowdowns, starter gas, and miscellaneous equipment associated with the compressor. Methane emissions from reciprocating compressor seals represent approximately 29% of the total emissions from reciprocating compressors. Note that the Eastern U.S. natural gas production did not include methane emissions from compressor seals in the reciprocating compressor emission factor, only emissions from the associated equipment (e.g., valves, connectors, and open-ended lines). Table 3-3 does include the estimated 129 gathering reciprocating compressors from the Eastern U.S. in the activity factor for natural gas production and estimates compressor seal methane emissions using the listed reciprocating compressor seal emission factor.

For natural gas processing, the total methane emissions from reciprocating and centrifugal compressors and associated equipment and operations were reported as 16,736,280 Mscf/yr (316,108 MT) and 5,626,500 Mscf/yr (106,271 MT), respectively. The methane emissions from reciprocating and centrifugal compressor seals represent 28% and 4.4%, respectively, of the total methane emissions from reciprocating and centrifugal compressors in the natural gas processing sector. In the natural gas transmission sector, the total methane emissions from reciprocating and centrifugal compressors were estimated to be 37,734,450 Mscf/yr (712,714 MT) and 7,559,100 Mscf/yr (142,773 MT), respectively. The methane emissions from reciprocating and centrifugal compressor seals represent 24% and 2.2%, respectively, of the total methane emissions from reciprocating and centrifugal compressors in the natural gas processing sector. The total methane emissions from reciprocating and centrifugal compressors and their associated equipment were estimated to be 10,763,160 Mscf/yr (203,290 MT) and 1,517,760 Mscf/yr (28,667 MT), respectively, for the natural gas storage sector. The methane emissions from reciprocating and centrifugal compressor seals represents 18% and 1.7%, respectively, of the total methane emissions from reciprocating and centrifugal compressors in the natural gas storage sector.

3.2   Natural Gas Industry Methane Emission Factor Improvement Study, Final Report (URS/UT, 2011)

The report describes the effort to update default methane emission factors for selected processes and equipment in the natural gas industry. These processes and equipment are believed to contribute the greatest uncertainty in the U.S. natural gas industry methane emissions inventory and concentrated on high emission rate leaks (fugitive leaks) from transmission, gathering/boosting, and gas processing reciprocating and centrifugal compressor components, including emissions from compressor vents (i.e., blowdown lines and compressor seals).

The emissions data were collected at 11 sites in Texas and New Mexico and included data from gathering and boosting stations, natural gas processing plants, and transmission stations. The sites were all constructed between the 1950s and 2000s. The total number of compressors that were measured included 66 reciprocating compressors and 18 centrifugal compressors, with 48 of the reciprocating compressors located at transmission compressor stations. For compressor seals, the measurements were conducted using the following steps:

- Where the reciprocating compressor rod packing vent lines were piped together (multiple cylinders joined into a single vent line for each compressor), the enclosed rotary vane anemometer was used to make the measurements at the top of the rod packing vent line;

- Where the reciprocating rod packing vent lines were individually vented to the atmosphere, each vent line was measured with a handheld hot wire anemometer; and

- For centrifugal compressors equipped with wet seals, measurements were made at the wet seal degassing fill port to the seal oil pump using plastic bags of known internal volume and measuring the required flow to fill the bag.

The study noted several technical issues with measuring emissions from a wet seal system including location of the flash emissions and configuration of the seal oil degassing system (which may include blowers or a flash drum/pot). The study noted that the wet seal measurements from this study should be used as a benchmark and requires further analysis before the measurements could be used to develop emission factors.

A summary of the testing results from the study are provided in Table 3-4. The study grouped the test results for centrifugal compressors located at natural gas gathering and boosting, processing and transmission together. The test data for reciprocating compressors were separated into units located at gathering and boosting stations and units located at transmission stations. The study found that the largest emission sources at a compressor stations are the compressor blowdown vent lines and the compressor seal vents (URS/UT, 2011).

14

Table 3-4. Sampling Results for Reciprocating and Centrifugal Compressor Seals

Compressor Vent Measured	Sample Size	Average Methane Emission Factor (Mscf/yr)	1996 GRI/EPA Emission Factor[a] (Mscf/yr)
Natural Gas Gathering/Boosting Reciprocating Compressors			
Average Rod Packing	15	241	9.48[b]
Natural Gas Transmission Reciprocating Compressors			
Average Rod Packing (Idle + depressurized)	5	12,236	396[c]
Average Rod Packing	2	29,603	
Natural Gas Gathering/Boosting, Processing and Transmission Centrifugal Compressors			
Average Wet Seal	9	8,137	396[d]

[a] (GRI/EPA, 1996b)
[b] Appendix B-4, assumes 4 seals per compressor.
[c] Table 4-15, adjusted for 79.1% time the compressor is pressurized.
[d] Table 4-15, adjusted for 24.2% time the compressor is pressurized.

The study authors concluded that the centrifugal compressor wet seal degassing vent emissions were much higher in comparison to the GRI/EPA emission factors. The study authors also determined that the average reciprocating compressor rod packing vent emissions that they calculated were significantly higher than the GRI/EPA study (GRI/EPA, 1996b).

3.3   Greenhouse Gas Reporting Program (U.S. EPA, 2013)

In October 2013, the EPA released 2012 greenhouse gas (GHG) data for Petroleum and Natural Gas Systems collected under the Greenhouse Gas Reporting Program (GHGRP). The GHGRP, which was required by Congress in the FY2008 Consolidated Appropriations Act, requires facilities to report data from large emission sources across a range of industry sectors, as well as suppliers of certain GHGs and products that would emit GHGs if released or combusted.

When reviewing this data and comparing it to other data sets or published literature, it is important to understand the GHGRP reporting requirements and the impacts of these requirements on the reported data. The GHGRP covers a subset of national emissions from Petroleum and Natural Gas Systems; a facility in the Petroleum and Natural Gas Systems source category is required to submit annual reports if total emissions are 25,000 metric tons carbon dioxide equivalent (CO_2e) or more. Facilities use uniform methods prescribed by the EPA to calculate GHG emissions, such as direct measurement, engineering calculations, or emission factors derived from direct measurement. In some cases, facilities have a choice of calculation methods for an emission source. Because some of the methods required direct measurement of emissions or parameters, for an interim period, the EPA made available the optional use of Best Available Monitoring Methods (BAMM) for unique or unusual circumstances. Where a facility used BAMM, it was required to follow emission calculations specified by the EPA, but was allowed to use alternative methods for determining inputs to calculate emissions.

Emissions for both reciprocating and centrifugal compressors are reported under the processing, transmission, underground gas storage, and liquid natural gas (LNG) import/export and storage segments. The calculation method varied by industry segment. Emissions from compressors in onshore production were calculated by using population counts multiplied by an emission factor. Emissions from compressors in the other industry segments were calculated by the use of direct measurement.

Table 3-4 shows activity data and emissions for reciprocating compressors for the natural gas processing, natural gas transmission, and underground natural gas storage industry segments. The EPA received data for 4,466 reciprocating compressors, including 2,149 reciprocating compressors in natural gas processing, 2,008 reciprocating compressors in natural gas transmission, and 309 reciprocating compressors in underground natural gas storage. Of the reciprocating compressors, BAMM was used to calculate emissions for 1,847 compressors, including 993 in natural gas processing, 790 in natural gas transmission, and 64 in underground natural gas storage.

Table 3-5. 2012 Direct Measurement Reported Process Emissions from Reciprocating Compressors from Natural Gas Processing, Natural Gas Transmission and Underground Natural Gas Storage

Industry Segment	Total Number of Reciprocating Compressors	Number of Reciprocating Compressors that used BAMM	Reported CH$_4$ Emissions (MT CO$_2$e)	Reported CH$_4$ Emissions[a] (MT)
Natural Gas Processing	2,149	993	1,009,045	48,050
Natural Gas Transmission	2,008	790	1,591,990	75,809
Underground Natural Gas Storage	309	64	160,809	7,658
Total	4,466	1,847	2,761,844	131,516

a. Conversion factors MT CO$_2$e to tons: 21 MT CH$_4$/MT CO$_2$e

Table 3-6 shows activity data and emissions for centrifugal compressors for the natural gas processing, natural gas transmission, and underground natural gas storage industry segments. For centrifugal compressors the number of compressors with wet seals is also shown. Overall emissions from centrifugal compressors were lower than those for reciprocating compressors, but the total number of reported compressors was lower as well. The EPA received data for 1,191 centrifugal compressors, including 428 centrifugal compressors in natural gas processing, 724 centrifugal compressors in natural gas transmission, and 39 centrifugal compressors in underground natural gas storage. Of these centrifugal compressors, BAMM was used to calculate emissions for 538 compressors, including 234 in natural gas processing, 292 in natural gas transmission, and 12 in underground natural gas storage.

Table 3-6. 2012 Direct Measurement Reported Process Emissions from Centrifugal Compressors from Natural Gas Processing, Natural Gas Transmission and Underground Natural Gas Storage

Industry Segment	Total Number of Centrifugal Compressors	Number of Centrifugal Compressors that used BAMM	Number of Centrifugal Compressors with Wet Seals	Reported CH_4 Emissions (MT CO_2e)	Reported CH_4 Emissions[a] (MT)
Natural Gas Processing	428	234	274	752,054	35,812
Natural Gas Transmission	724	292	291	439,714	20,939
Underground Natural Gas Storage	39	12	23	118,500	5,643
Total	1,191	538	588	1,310,268	62,394

a.   Conversion factors: 21 MT CH_4/MT CO_2e

3.4   Inventory of U.S. Greenhouse Gas Emissions and Sinks: 1990-2012 (U.S. EPA, 2014)

The EPA leads the development of the annual Inventory of U.S. Greenhouse Gas Emissions and Sinks (GHG Inventory). This report tracks total U.S. GHG emissions and removals by source and by economic sector over a time series, beginning with 1990. The U.S. submits the GHG Inventory to the United Nations Framework Convention on Climate Change (UNFCCC) as an annual reporting requirement. The GHG Inventory includes estimates of methane and carbon dioxide for natural gas systems (production through distribution) and petroleum systems (production through refining).

The 2014 GHG Inventory (published in 2014; containing emissions data for 1990-2012) calculates net methane emissions for reciprocating compressors using emission factors based on the GRI/EPA study (GRI/U.S. EPA, 1996a). The factors are used to develop potential emissions. The total potential emissions are reduced by known controls or practices that reduce emissions to calculate net emissions. For centrifugal compressors, the EPA has developed emission factors for both wet seal and dry seal compressors that are used to directly calculate net emissions (i.e., after control).

For the natural gas production stage, emission factors for gathering compressors are regional and cover small and large reciprocating compressors (no centrifugal compressors).

For natural gas processing, and transmission and storage, the emission factors are for reciprocating compressors and the two types of centrifugal compressors (wet and dry seal). For LNG storage and import/export, there are factors for reciprocating and centrifugal compressors. The emission factors used to calculate methane emission for compressors for the 2014 GHG Inventory are summarized in Table 3-8.

Table 3-9 summarizes the activity data and 2012 calculated potential methane emissions for compressors by industry segment and compressor type.

Table 3-8. Natural Gas Sector Methane Emission Factors for Compressors

| | Emission Factor (scf/day/compressor) | | | |
| | Reciprocating | | Centrifugal | |
Industry Activity	**Small[1]**	**Large**	**Wet Seal**	**Dry Seal**
Production	263-312	14,947-17,728	-	
Processing	11,196		51,370	25,189
Transmission	15,205		50,222	32,208
Storage	21,116		45,441	31,989
LNG Storage/Import	21,116		30,573	

---

[1] The GRI/EPA study defines small gathering compressors as compressors on the overhead lines from gas well separators and associated gas well separators. Large gathering compressors are compressors at large gathering compressor stations (stations with **8** compressors or more).

Table 3-9. Summary of Natural Gas Sector Compressor Activity and Calculated Potential Methane Emissions

Industry Segment	Activity (Compressor Units)	Calculated Potential Methane Emissions (MT)
Production		
Reciprocating (small)	35,930	70,859
Reciprocating (large)	136	15,400
Processing		
Reciprocating	5,624	442,634
Centrifugal (wet seal)	658	237,724
Centrifugal (dry seal)	248	43,937
Transmission		
Reciprocating	7,235	773,294
Centrifugal (wet seal)	659	232,826
Centrifugal (dry seal)	66	14,972
Storage		
Reciprocating	1,012	150,225
Centrifugal (wet seal)	70	22,347
Centrifugal (dry seal)	29	6,532
LNG Storage		
Reciprocating	270	40,147
Centrifugal	64	13,766
LNG Import Terminal		
Reciprocating	37	5,552
Centrifugal	7	1,419

The GHG Inventory emissions calculations used regional values by industry segment for the methane content in natural gas. The average national value for general sources was 83.3% methane for 2012.

The net 2012 methane emissions reported for compressors for the 2014 GHG Inventory were 86,259 MT for the natural gas production segment, 724,295 MT for the natural gas processing segment, and 1,261,080 MT for the natural gas transmission and storage segment, for a total of 2,071,633 MT of methane.

3.5 Development of the New Source Performance Standard (NSPS) For Oil and Natural Gas Production (U.S. EPA, 2011b and U.S. EPA, 2012a)[2]

VOC emission factors were developed for reciprocating and centrifugal compressors in order to support the development of subpart OOOO. In order to develop these factors the EPA used information from the GHGRP[3], the GHG Inventory, the EPA's Natural Gas STAR Program, and a study by the GRI/EPA study. Updates to the GHGRP and the GHG Inventory have occurred since this analysis, however, it is presented here for completeness.

The methodology for estimating emissions from reciprocating compressor rod packing was to use the methane emission factors referenced in the EPA/GRI study (GRI/EPA, 1996a) and use the methane-to-pollutant ratios developed in the gas composition memorandum developed for subpart OOOO. (EC/R, 2011) The emission factors in the EPA/GRI study were expressed in thousand standard cubic feet per cylinder (Mscf/cyl), and were multiplied by the average number of cylinders per reciprocating compressor at each oil and gas industry segment. The volumetric methane emission rate was converted to a mass emission rate using a density of 41.63 pounds of methane per thousand cubic feet. This conversion factor was developed assuming that methane is an ideal gas and using the ideal gas law to calculate the density.

The centrifugal compressor emission factors for wet seals and dry seals were based on emission factors from the 2012 GHG Inventory (published in 2012; containing emissions data for 1990-2010). The wet seal methane emission factor was calculated based on a sampling of 48 wet seal centrifugal compressors. The dry seal methane emission factor was based on data collected by the Natural Gas STAR Program. The methane emissions were converted to VOC emissions using the same gas composition ratios that were used for reciprocating engines. (EC/R, 2011) A summary of the methane emission factors is presented in Table 3-10.

---

[2] Unless otherwise indicated, the following sections are excerpts from either Section 6 of the technical support document for the proposed subpart OOOO (U.S. EPA, 2011b) or Section 6.0 of the technical support document for the final subpart OOOO rule (U.S. EPA, 2012a).

[3] http://www.epa.gov/ghgreporting/

Table 3-10. Methane Emission Factors for Reciprocating and Centrifugal Compressors

Oil and Gas Industry Segment	Reciprocating Compressors			Centrifugal Compressors	
	Methane Emission Factor (scf/hr-cylinder)	Average Number of Cylinders per Compressor	Pressurized Factor (% of hour/year Compressor Pressurized)	Wet Seal Methane Emission Factor (scf/minute)	Dry Seals Methane Emission Factor (scf/minute)
Production (Well Pads)	0.271[a]	4	100%	N/A[f]	N/A[f]
Gathering & Boosting	25.9[b]	3.3	79.1%	N/A[f]	N/A[f]
Processing	57[c]	2.5	89.7%	47.7[g]	6[g]
Transmission	57[d]	3.3	79.1%	47.7[g]	6[g]
Storage	51[e]	4.5	67.5%	47.7[g]	6[g]

[a] (GRI/EPA, 1996a), Table 4-8.
[b] Clearstone Engineering Ltd. *Cost-Effective Directed Inspection and Maintenance Control Opportunities at Five Gas Processing Plants and Upstream Gathering Compressor Stations and Well Sites.*: 2006.
[c] (GRI/EPA, 1996a), Table 4-14.
[d] (GRI/EPA, 1996a) Table 4-17.
[e] (GRI/EPA, 1996a) Table 4-24.
[f] The 1996 EPA/GRI Study Volume 11[4], does not report any centrifugal compressors in the production or gathering/boosting sectors; therefore, no emission factor data were published for those two sectors.
[g] (U.S. EPA, 2011a), Annex 3. Page A-153.
Source: Derived from (U.S. EPA, 2011b), Table 6-2 and (U.S. EPA, 2012a), Table 6-1

Once the methane emission rates for compressors were calculated using the emission factors, ratios were applied to the methane emissions to estimate VOC emissions. The specific ratios that were used for this analysis were 0.278 pounds VOC per pound of methane for the production and processing segments, and 0.0277 pounds VOC per pound of methane for the transmission and storage segments. A summary of the baseline individual compressor emission rates are shown in Table 3-11 for each of the oil and gas industry segments.

---

[4]   EPA/GRI (1996) Methane Emission from the Natural Gas Industry, Vol. 11, .Pages 11 – 15. Available at: http://epa.gov/gasstar/documents/emissions_report/11_compressor.pdf

Table 3-11. Baseline Emission Rates for Reciprocating and Centrifugal Compressors

Industry Segment/ Compressor Type	Baseline Emission Estimates (tons/compressor/year)	
	Methane	VOC
Reciprocating Compressors		
Production (Well Pads)	0.198	0.0549
Gathering & Boosting	12.3	3.42
Processing	23.3	6.48
Transmission	27.1	0.751
Storage	28.2	0.782
Centrifugal Compressors (Wet seals)		
Processing	228	20.5
Transmission	126	3.50
Storage	126	3.50
Centrifugal Compressors (Dry seals)		
Processing	28.6	2.58
Transmission	15.9	0.440
Storage	15.9	0.440

Source: Derived from (U.S. EPA, 2011b), Table 6-2 and (U.S. EPA, 2012a), Table 6-1

The analysis performed in the technical support document (TSD) to proposed subpart OOOO (U.S. EPA, 2011b) was designed to provide information about new compressors for the purposes of establishing new source performance standards; accordingly, the analysis did not estimate nationwide emissions for all compressors.

3.6 Characterizing Pivotal Methane Emissions from the Oil and Natural Gas Sector, (API /ANGA, 2012)

The API/ANGA study (API/ANGA, 2012) is an analysis of industry survey data that includes data from over 20 companies covering over 90,000 gas wells. This study sample population includes representation from most of the geographic regions of the country as well as most of the geologic formations currently developed by the industry.

With respect to compressors, the API/ANGA study collected information related to the activity count for centrifugal compressors, specifically to supplement the EPA's data on the prevalence of wet seal and dry seal compressors in the industry. According to the survey results, the data collected represented approximately 5% of the national centrifugal compressor count for gas processing operations (38 centrifugal compressors from the survey, compared to 811 from 2012 GHG Inventory). Of the gas processing centrifugal compressors reported through the survey, 79% were dry seal compressors and 21% were wet seal units. If the results of the survey were considered to be representative, the authors assert that the EPA's current ratio of 80% wet seal and 20% dry seals severely overestimates the emissions from the wet seal compressors. Based on the emission factors from Table A-123 of Annex 3 of the 2012 GHG Inventory, the methane emissions from centrifugal compressors would be 190,573 tons (172,887 MT) compared to 288,068 tons (261,334 MT) from the 2012 GHG Inventory. This would equate to an approximate 34% reduction in the emissions from this source. The authors recommended using the GHGRP data to further refine these activity numbers.

With respect to production and gathering facilities that use centrifugal compressors, the API/ANGA survey responses reported only 550 centrifugal compressors associated with production and gathering at 21 participating companies. The authors noted that the 2012 GHG Inventory did not include centrifugal compressors in production/gathering operations. The study reported that, on a well basis, the survey response equates to 0.07 centrifugal compressor per gas well with 81% of those being dry seal and the remaining being wet seal. The authors recommended that the EPA continue to refine these numbers using data from the GHGRP.

3.7 Economic Analysis of Methane Emission Reduction Opportunities in the U.S. Onshore Oil and Natural Gas Industries (ICF International, 2014)

The Environmental Defense Fund (EDF) commissioned ICF International (ICF) to conduct an economic analysis of methane emission reduction opportunities from the oil and natural gas industry to identify the most cost-effective approach to reduce methane emissions from the industry. The study projects the estimated growth of methane emissions through 2018 and focuses its economic analysis on 22 methane emission sources in the oil and natural gas industry (referred to as the targeted emission sources). These targeted emission sources represent 80% of the study's projected 2018 methane emissions from onshore oil and gas industry sources. Centrifugal compressor and reciprocating compressor emission sources were included in their list of targeted emission sources.

The study relied on the 2013 GHG Inventory for methane emissions data for the oil and natural gas sector. These emissions data for compressors were revised to include updated information from the GHGRP, data from the 1996 GRI/EPA study of methane emissions, information on the Federal Energy Regulatory Commission (FERC) website, data obtained from the U.S. Department of Transportation Pipeline and Hazardous Materials Safety Administration and information from various state energy and environmental departments. The revised ICF 2011 baseline methane emissions estimates were then used as the basis for projecting onshore methane emissions to 2018. A summary of the most significant revisions made to the 2013 GHG Inventory activity and emission factors to develop the revised ICF 2011 baseline by industry segment are presented in Section 3.8.1. The methodology used to project onshore methane emissions from the revised 2013 GHG Inventory (referred to as the ICF 2011 baseline) to 2018 for compressors is presented in Section 3.8.2.

3.7.1 ICF 2011 Baseline

The ICF study breaks out emissions by natural gas segment (gas production, gathering and boosting, gas processing, gas transmission, gas storage, LNG and gas distribution) and petroleum segment (oil production, oil transportation and oil refining). The most significant revisions made to the 2013 GHG Inventory to develop the ICF 2011 baseline for compressors are summarized by industry segment in the following paragraphs. Note that no emission factor or activity changes related to compressors were made for the gas production, oil production, oil transportation, and oil refining segments.

3.7.1.1 Gathering and Boosting Segment

Reciprocating Compressors

- Updated the 1996 EPA/GRI study emissions factors used in the 2013 GHG Inventory using information obtained from five state energy agencies (Texas, Colorado, Wyoming, Oklahoma and Pennsylvania) on permitted engines for production and gathering compressors in the petroleum and natural gas industry. These data were split into large and small compressors using the 1,600 horsepower (hp) threshold from the 1996 EPA/GRI study. The state data showed a larger percentage of large compressors than assumed in the 2013 GHG Inventory. A new weighted average factor was calculated using the 1996 EPA/GRI study emission factors. The new methane emission factor for all gathering compressors was calculated at 1,980 scf/day/compressor.
- The reciprocating compressor emission factor used in the 2013 GHG Inventory was updated to distinguish compressor seal emissions versus compressor fugitives (which are combined in the GHG Inventory) using the 1996 EPA/GRI study emission factors, whereby compressors seals were then separated into two categories: reciprocating compressors – non-seals (75%) and reciprocating compressors – seals (25%).
- Developed new activity factors for reciprocating compressors using information obtained from the five state energy agencies (discussed above) by using the 2013 GHG Inventory ratio of compressors in these five states to the national count of compressors to obtain a new national reciprocating compressor count of 15,687.

Based on these revisions, ICF estimated the net change in methane emissions from reciprocating compressors (as compared with the 2013 GHG Inventory) to be 166% or an increase to 11 Bcf (228,965 tons).

<u>Centrifugal Compressors</u>

- Created a new emission category for wet seal centrifugal compressors based on information obtained from the GHGRP that included 162 wet seal centrifugal compressors used in the upstream sector. ICF assumed that the respondents under the GHGRP represented 85% of the industry. Therefore, ICF adjusted the number of wet seal centrifugal compressors to be 191. ICF used an emission factor of 12,000,000 scf/year/compressor (from subpart W) and their estimated number of wet seal centrifugal compressors to estimate methane emissions for the 2011 baseline (over 2 Bcf [41,630 tons] methane).

3.7.1.2 Gas Processing

- Reciprocating compressors emission factor updated to breakout emissions from compressor seals versus "other" compressor fugitives as discussed in Section 3.8.1.1.

3.7.1.3 Gas Transmission

- Number of compressor stations revised from 1,808 to 1,768 (based on a change in pipeline miles included in the 2013 GHG Inventory using data obtained from the U.S. Department of Transportation Pipeline and Hazardous Materials Safety Administration indicating a lower value for transmission pipeline miles), resulting in an emissions decrease of just over 2%.

- Number of reciprocating compressors changed from 7,270 to 7,111 (based on changes to the pipeline miles included in the 2013 GHG Inventory – see above), resulting in an emissions decrease of over 2%.

- Number of centrifugal compressors revised from 654 to 648 (based on changes to the pipeline miles included in the 2013 GHG Inventory – see above), resulting in an emissions decrease of over 2%.

3.7.1.4 Gas Storage

- Reciprocating compressors emission factor updated to breakout emissions from compressor seals versus "other" compressor fugitives as discussed in Section 3.8.1.1.

3.7.2   ICF Projections to 2018

Emissions projections are not the subject of this paper; therefore, the estimates of 2018 emissions produced in the ICF study are not presented here. However, the ICF study uses the projections to evaluate emissions mitigation techniques for compressors, which are addressed in this paper. Those mitigation techniques are discussed in detail in Section 4 of this paper. The methodology the ICF study used to project emissions to 2018 is described here in order to provide context for the later discussion of mitigation techniques.

The primary sources used for projecting onshore methane emissions for centrifugal and reciprocating compressors for 2018 included the INGAA Foundation *North American Midstream Infrastructure Through 2035-A Secure Energy Future* report (ICF, 2011), an analysis of past projected infrastructure change, FERC and ICF information on emission reductions anticipated as a result of regulation (40 CFR Part 60, subpart OOOO).

The INGAA report provided yearly forecast information of incremental gathering pipeline miles, gas processing plants and processing compressor counts that were used with existing activity data from the 2013 GHG Inventory to estimate a regional activity factor for use to make projections out to 2018. For the gathering and boosting segment, the activity factors were estimated based on a ratio of pipeline miles in 2018 to pipeline miles in the ICF 2011 baseline to obtain 2018 activity levels. For the gas processing segment, the activity factors were estimated based on a ratio between the compressor count in 2018 and the compressor count in the ICF 2011 baseline. For the gas transmission segment, projections out to 2018 were based on an analysis of past pipeline infrastructure changes, where the change in the length of transmission pipeline from 1990 to 2011 was used to establish an incremental value based on trends that were then used to project the pipeline miles for 2018.

The new 2018 forecast of emissions for centrifugal and reciprocating compressors (for all but production and transmission segments) were adjusted to account for emission reductions that are expected as a result of the EPA's NSPS, subpart OOOO.

Further information included in this study on mitigation or emission reduction options, methane control costs, and their estimates for the potential for VOC emissions co-control benefits from their use is presented in Section 4 of this document.

4.0 AVAILABLE COMPRESSOR EMISSIONS MITIGATION TECHNIQUES

Emissions mitigation options for reciprocating compressors involve techniques that limit the leaking of natural gas past the piston rod packing, including replacement of the compressor rod packing, replacement of the piston rod, and the refitting or realignment of the piston rod. The EPA is also aware of new technologies that enable the emissions to be captured and either routed to a combustion device or a useful process. Emission mitigation options for centrifugal compressors limit the leaking of natural gas across the rotating shaft using a mechanical dry seal, or capture the gas and route it to a useful process or to a combustion device. A discussion of these techniques and their costs is presented in the following sections.

4.1 Reciprocating Compressor - Rod Packing Replacement

4.1.1 Description

The potential emission reduction options for reciprocating compressors include control techniques that limit the leaking of natural gas past the piston rod packing. Reciprocating compressor rod packing consists of a series of flexible rings that fit around a shaft to create a seal against leakage. Rod packing emissions typically occur around the rings from slight movement of the rings in the cups as the rod moves, but can also occur through the "nose gasket" around the packing case, between the packing cups, and between the rings and shaft (see Figure 2-1). Mitigation options for these emissions include replacement of the compressor rod packing, replacement of the piston rod, and the refitting or realignment of the piston rod (U.S. EPA, 2006a).

The replacement of the rod packing is a maintenance task performed on reciprocating compressors to reduce the leakage of natural gas past the piston rod. Over time, the packing rings wear and allow more natural gas to escape around the piston rod. Regular replacement of these rings reduces VOC and methane emissions.

Like the packing rings, piston rods on reciprocating compressors also deteriorate. Rods can wear "out-of-round" or taper when poorly aligned, which affects the fit of packing rings against the shaft (and therefore the tightness of the seal) and the rate of ring wear. Replacing or upgrading the rod can reduce reciprocating compressor rod packing emissions. Also, upgrading piston rods by coating them with tungsten carbide or chrome reduces wear over the life of the rod (U.S. EPA, 2006a).

4.1.2   Effectiveness

As discussed above, regular replacement of the reciprocating compressor rod packing can reduce the leaking of natural gas across the piston rod. The emission reductions are related to the rate of deterioration and the frequency of replacement.

Subpart OOOO Technical Support Document (U.S. EPA, 2011b)

In the TSD for the subpart OOOO rulemaking, the expected emission reductions from a rod packing replacement were calculated by comparing the average rod packing emissions with the average emissions from newly installed and worn-in rod packing (U.S. EPA, 2011b). For gathering and boosting compressors, the analysis calculated the potential methane emission reductions by multiplying the number of new reciprocating compressors by the difference between the average rod packing emission factor in Table 3-10 by the average emission factor from a newly installed rod packing. The average rod packing emission factor used for gathering and boosting compressors was developed from the Clearstone II study (Clearstone, 2006) using rod packing measurement data (which was adjusted for the percent of time transmission compressors are operating) (GRI/U.S. EPA, 1996). For wellhead reciprocating compressors, the analysis calculated a percentage reduction using the transmission emission factor from the 1996 EPA/GRI report and the minimum emissions rate from a newly installed rod packing to determine methane emission reductions. The emission reductions for the processing, transmission, and storage segments were calculated by multiplying the number of new reciprocating compressors in each segment and the difference between the average rod packing emission factors in Table 3-10 (GRI/U.S. EPA, 1996) and the average emission factor from newly installed rod packing. Newly installed packing average methane emissions were assumed to be 11.5 cubic feet per hour per cylinder (U.S. EPA, 2006a).

A summary of the estimated emission reductions for reciprocating rod packing replacement for each of the oil and gas segments from the subpart OOOO TSD is shown in Table 4-1. The emissions of VOC were calculated using the methane emission reductions calculated above and the gas composition (EC/R, 2011) for each of the segments.

Table 4-1. Estimated Annual Individual and Nationwide Emission Reductions from Replacing Rod Packing in Reciprocating Compressors

Oil & Gas Segment	Individual Compressor Emission Reductions (tons/compressor-year)	
	Methane	**VOC**
Production (Well Pads)	0.158	0.0439
Gathering & Boosting	6.84	1.90
Processing	18.6	5.18
Transmission	21.7	0.600
Storage	21.8	0.604

Economic Rod Packing Replacement

The Natural Gas STAR Lessons Learned document titled "Reducing Methane Emissions from Compressor Rod Packing Systems" (U.S. EPA, 2006a) states that a new, properly installed rod packing system should leak approximately 11 to 12 standard cubic feet per hour (scfh) of gas. The effectiveness of the system on minimizing leaks is reliant on the fit of, and wear to the rod packing components (such as the rod packing material, the cups that hold it, and the piston rod). As the rod packing system ages, the leak rates will increase. Eventually, the leak rate will reach a point where the amount of gas saved by replacing the rod packing will justify the cost of performing the replacement. In some cases, the economic threshold for replacement can be as low as 30 scfh of gas leakage. However, if the rod packing systems are not well maintained, the leakage rates can far exceed that value. In one instance, a Natural Gas STAR partner reported emissions from an aging rod packing system to be as high as 900 scfh.

Updated rod packing components made from newer materials can also help improve the life and performance of the rod packing system. Another potential option is replacing the bronze metallic rod packing rings with longer lasting carbon-impregnated Teflon rings. Compressor rods can also be coated with chrome or tungsten carbide to reduce wear and extend the life of the piston rod (U.S. EPA, 2006a).

4.1.3   Cost of Controls

The Natural Gas STAR Lessons Learned document estimates the cost to replace the packing rings on reciprocating compressors to be $1,620 per cylinder. The replacement of rod packing for reciprocating compressors occurs on average every four years based on industry information from the Natural Gas STAR Program. (U.S. EPA, 2006a)

The TSD for the subpart OOOO rulemaking used the above costs from the Natural Gas STAR Lessons Learned document and operating factors from the GRI/EPA study to determine the costs and gas savings from rod packing replacement (U.S. EPA, 2011b). The weighted hours, on average, per year the reciprocating compressor is pressurized was calculated to be 98.9%

using the operating factors presented in Table 3-2 of this paper (GRI/EPA, 1996a). The calculated years were assumed to be the equipment life of the compressor rod packing. Table 3-2 was used to estimate the average number of cylinders per compressor for each industry segment. Information reviewed did not identify any annual or periodic maintenance costs for the rod packing systems. Because replacement of rod packing systems reduces gas emissions, a monetary savings can be realized that is associated with the amount of gas saved with reciprocating compressor rod packing replacement. The savings were estimated using a natural gas price of $4.00 per Mcf (U.S. EIA, 2010). This gas price was used to calculate the annual savings using the methane emission reductions in Table 4-1. The savings over the useful equipment life of the rod packing system was then calculated based on equipment life discussed above. A summary of the estimated capital costs and estimated gas savings for each of the oil and gas segments is shown in Table 4-2.

Table 4-2. Capital Cost and Gas Savings for Reciprocating Compressor Rod Packing Replacement

Oil and Gas Segment	Capital Cost per compressor ($2008)	Gas Savings for Equipment Life per Compressor
Production	$6,480	$2,493
Gathering & Boosting	$5,346	$1,669
Processing	$4,050	$1,413
Transmission	$5,346	$1,669
Storage	$7,290	$2,276

The ICF International study (ICF, 2014) evaluated the effectiveness of replacing rod packing systems in reciprocating compressors for existing sources to reduce methane emissions assuming 98% control with timely replacement to minimize emissions. Their analysis assumed capital costs of $2,000 every 3 years for replacement of the packing system, and revenue benefits from reduction of methane emission losses of $3,500 (at $4.00/Mcf gas). The estimated payback period for this control option was estimated to be seven months. National emission reductions were estimated to be 3.6 Bcf (74,934 tons) methane/yr. ICF estimated national annualized costs of replacing rod packing systems to be $22.3 million/yr and total initial capital costs to be an estimated $182.3 million. ICF also estimated that VOC emissions would be reduced by 8 kilotons (or approximately 8,816 tons) at a cost of $2,784/ton of VOC reduced. ICF concluded that replacing rod packing systems in reciprocating compressors can significantly reduce methane emissions and increase savings.

4.2   Reciprocating Compressor – Gas Recovery

4.2.1   Description

The potential emission reduction options for reciprocating compressors include control techniques that recover natural gas leaking past the piston rod packing. The EPA is aware of one company, REM Technology, Inc., that has developed a system that captures the gas that would otherwise be vented and routes it back to the compressor engine to be used as fuel (REM, 2012). The vent gases are passed through a valve train that includes a demister and then are injected into the engine intake air after the air filter. The EPA is aware that this technology has been deployed commercially, but does not have any information on the extent it is used in the field.

Another method for capturing emissions from reciprocating compressor rod packing vents is to manifold the vent line to a vapor recovery unit (VRU) system. A VRU is a simple system designed to capture vented gas streams, usually from tanks, that would otherwise go to the atmosphere. The main components of the system include a compressor and scrubber. If a VRU system is already in place at a facility with reciprocating compressors, it is often possible to route the vent streams to tanks, allowing the vented rod packing gas to be picked up by the VRU. The recovered gas can then be sold or routed for fuel or other meaningful use onsite. If the gas cannot be used productively, it can also be sent to a flare system. While flaring may have a higher cost than venting to the atmosphere, this practice can reduce methane and VOC emissions.

4.2.2 Effectiveness

REM Technology estimates that the gas recovery system can result in the elimination of over 99% of VOC and methane emissions that would otherwise occur from the venting of the emissions from the compressor rod packing (REM, 2013). The emissions that would have been vented are combusted in the compressor engine to generate power. This technique is discussed further in the Natural Gas STAR PRO Fact Sheet titled "Install Automated Air/Fuel Ratio Controls" (U.S. EPA, 2011c). This document reported an average fuel gas savings of 78 thousand cubic feet per day (Mcfd) per engine with the gas recovery system installed.

If the facility is able to route rod packing vents to a VRU system, it is possible to recover approximately 95-100% of emissions. If the gas is routed the gas to a flare, approximately 95% of the methane and VOCs are reduced.

4.2.3 Cost of Controls

The EPA has not been able to obtain cost data on the REM technology. Some costs would be mitigated by fuel gas savings, as using the captured gas to displace some of the purchased fuel would require less fuel to be purchased in order to run the compressor engine.

For a VRU, assuming the proper equipment is already available at the facility, capturing the rod packing gas would require minimal costs. The investment would only need to include the cost of piping and installation. While the EPA has not obtained a cost estimate specifically for routing rod packing vents to a VRU, this process has been studied for dehydrators and would be similar for rod packing systems. According to the Natural Gas STAR PRO Fact Sheet titled "Pipe Glycol Dehydrator to Vapor Recovery Unit" (U.S. EPA, 2011d), the cost for planning and installing additional piping is approximately $2,000. Routing to a VRU also provides additional incentive as there is a value associated with recovered gas. However, the installation of a VRU to only capture rod packing emissions may not be economically viable if an additional compressor system is required. If the VRU is already present at the facility, the incremental cost to capture the rod packing vent gas can be recovered from the value of the additional captured gas.

4.3   Centrifugal Compressor - Dry Seals

4.3.1   Description

Centrifugal compressor dry seals operate mechanically under the opposing force created by hydrodynamic grooves and springs. The hydrodynamic grooves are etched into the surface of the rotating ring affixed to the compressor shaft. When the compressor is not rotating, the stationary ring in the seal housing is pressed against the rotating ring by springs. When the compressor shaft rotates at high speed, compressed gas has only one pathway to leak down the shaft, and that is between the rotating and stationary rings. This gas is pumped between the rings by grooves in the rotating ring. The opposing force of high-pressure gas pumped between the rings and springs trying to push the rings together creates a very thin gap between the rings through which little gas can leak (see Figure 2-3). While the compressor is operating, the rings are not in contact with each other; therefore, they do not wear or need lubrication. O-rings seal the stationary rings in the seal case.

Dry seals reduce emissions and, at the same time, they reduce operating costs and enhance compressor efficiency. Economic and environmental benefits of dry seals include:

- Gas Leak Rates. Wet seals generate vented emissions during degassing of the circulating oil. Gas separated from the seal oil before the oil is recirculated is usually vented to the atmosphere, bringing the total leakage rate for tandem wet seals to 47.7 scfm natural gas per compressor (U.S. EPA/ICR, 2009) (U.S. EPA, 2011a, Annex 3, page A-153).
- Mechanically Simpler. Dry seal systems do not require additional oil circulation components and treatment facilities.
- Reduced Power Consumption. Because dry seals have no accessory oil circulation pumps and systems, they avoid "parasitic" equipment power losses. Wet seal systems require 50 to 100 kW per hour, while dry seal systems need about 5 kW of power per hour.
- Improved Reliability. The highest percentage of downtime for a compressor using wet seals is due to seal system problems. Dry seals have fewer ancillary components, which translates into higher overall reliability and less compressor downtime.

- Lower Maintenance. Dry seal systems have lower maintenance costs than wet seals because they do not have moving parts associated with oil circulation (e.g., pumps, control valves, relief valves, and the seal oil cost itself).

- Elimination of Oil Leakage from Wet Seals. Substituting dry seals for wet seals eliminates seal oil leakage into the pipeline, thus avoiding contamination of the gas and degradation of the pipeline.

4.3.2 Effectiveness

The emissions reduction effectiveness of the dry seals was calculated in the TSD for the proposed subpart OOOO (U.S. EPA, 2011b) by subtracting the dry seal emissions from a centrifugal compressor equipped with wet seals. The centrifugal compressor emission factors in Table 3-2 were used in combination with an operating factor of 43.6% for processing centrifugal compressors and 24.2% for transmission centrifugal compressors. The operating factors are used to account for the percent of time in a year that a compressor is in the operating mode. The operating factors for the processing and transmission sectors are based on data in the EPA/GRI study (GRI/EPA, 1996a). The wet seals emission factor is an average of 48 different wet seal centrifugal compressors. The dry seal emission factor is based on information from the Natural Gas STAR Program (U.S. EPA, 2006b). A summary of the emission reduction from the replacement of wet seals with dry seals is shown in Table 4-3.

Table 4-3. Estimated Annual Centrifugal Compressor Emission Reductions from Replacing Wet Seals with Dry Seals

Oil & Gas Segment	Individual Compressor Emission Reductions (ton/compressor-year)	
	Methane	VOC
Processing	199	18.0
Transmission/Storage	110	3.06

4.3.3 Cost of Controls

The price difference between a brand new dry seal and brand new wet seal centrifugal compressor is small relative to the cost for the entire compressor. The analysis in the TSD for proposed subpart OOOO assumed the additional capital cost for a dry seal compressor to be $75,000, with an equipment life of 10 years (U.S. EPA, 2011b).

The Natural Gas STAR Program estimated that the operation and maintenance savings from the installation of dry seals is $88,300 annually in comparison to wet seals (U.S. EPA, 2006b). Monetary savings associated with the amount of gas saved with the replacement of wet seals with dry seals for centrifugal compressors was estimated using a natural gas price of $4.00 per Mcf (U.S. EIA, 2010). This cost was used to calculate the annual gas savings using the methane emission reductions in Table 4-2. There is no gas savings cost benefits for transmission and storage facilities, because it is assumed the owners of the compressor station do not own the natural gas that is compressed at the station. A summary of the capital cost, annual operation and maintenance cost and the natural gas savings for replacing wet seals with dry seals is presented in Table 4-4. As shown in the table, there is a net savings after one year of operation without considering any potential natural gas savings.

Table 4-4. Costs for Replacing Centrifugal Compressor Wet Seals with Dry Seals

Oil and Gas Segment	Capital Cost per compressor ($2008)	Annual Operation and Maintenance Savings ($/compressor)	Annual Natural Gas Savings ($/compressor)
Processing	$75,000	$88,300	$46,109
Transmission/Storage	$75,000	$88,300	0

The ICF International study (ICF, 2014) evaluated replacing a wet seal with a dry seal for centrifugal compressors (assuming 97% control of methane emissions) as a control option using their 2018 projected methane emission estimates (discussed in Section 3.8.2 of this document). Their analysis assumed retrofit capital and annual operating costs of $400,000 and $17,500, respectively, and annual product revenue benefits of $180,500 (assuming $4/Mcf of gas) due to the reduction of product loss to the atmosphere. The report states that a dry seal retrofit is not common due to the high up-front costs and the downtime that would be required, and estimates that the payback period would be 29 months. The report also states that information from vendors indicates that 90% of new centrifugal compressors are already equipped with dry seals.

4.4   Centrifugal Compressor - Wet Seal with a Flare

4.4.1   Description

Another emission reduction option for centrifugal compressors equipped with wet seals is to route the emissions to a combustion device or capture the emissions and route them to a fuel system. A wet seal system uses oil that is circulated under high pressure between three rings around the compressor shaft, forming a barrier against the compressed gas. The center ring is attached to the rotating shaft, while the two rings on each side are stationary in the seal housing, pressed against a thin film of oil flowing between the rings to both lubricate and act as a leak barrier. Compressed gas becomes absorbed and entrained in the fluid barrier and is removed using a heater, flash tank, or other degassing technique so that the oil can be recirculated back to the wet seal. The removed gas is either combusted, released to the atmosphere, or captured and routed to a process. The emission reduction technique investigated in this section is the use of wet seals with the removed gas sent to an enclosed flare.

4.4.2   Effectiveness

Flares have been used in the oil and gas industry to combust gas streams that have VOC and methane constituents. A flare typically achieves 95% reduction of these compounds when operated according to the manufacturer instructions. For this analysis, it was assumed that 100% of the entrained gas from the seal oil that is removed in the degassing process would be directed to a flare that achieves 95% reduction of organic compounds. The wet seal emissions in Table 3-2 were used along with the control efficiency of the flare to calculate the emissions reductions from this option. A summary of the emission reductions is presented in Table 4-5.

Table 4-5. Estimated Annual Centrifugal Compressor Emission Reductions from Wet Seals Routed to a Flare

Oil & Gas Segment	Individual Compressor Emission Reductions (tons/compressor-year)	
	Methane	VOC
Processing	216	19.5
Transmission/Storage	120	3.32

4.4.3 Cost of Controls

The capital and annual costs of the enclosed flare were calculated using the methodology in the EPA Control Cost Manual. (U.S. EPA, Cost) The heat content of the gas stream was calculated using information from an the EPA study to estimate the composition of natural gas previously developed for the analysis of subpart OOOO. (EC/R, 2011) A summary of the capital and annual operation and maintenance costs for wet seals routed to a flare is presented in Table 4-6. There is no cost saving estimated for this option because the recovered gas is combusted.

Table 4-6. Costs for Centrifugal Compressor Wet Seals Routed to a Flare

Oil and Gas Segment	Capital Cost ($2008)	Annual Cost per Compressor
Processing	$67,918	$98, 329
Transmission/Storage	$67,918	$98,329

4.5 Centrifugal Compressor - Wet Seals with Gas Recovery for Use

4.5.1 Description

The final option for emissions reduction for wet seal centrifugal compressors is to capture and reroute the emissions back into the process. Based on comments received during development of subpart OOOO, in some cases gas may be routed back to the compressor suction or fuel system.

The emissions reductions for wet seal centrifugal compressors in the processing sector and transmission and storage sectors are summarized in Table 4-7 using 95% control efficiency for the capture system.

**Table 4-7. Wet Seal Centrifugal Compressor Emission Reductions
at 95% Capture and Control**

Source	VOC (tpy)	Methane (tpy)
Emissions Reductions Per Wet Seal Centrifugal Compressor – Natural Gas Processing	19.5	216
Emissions Reductions Per Wet Seal Centrifugal Compressor – Transmission/Storage	3.32	120

4.5.3   Cost of Controls

Natural Gas STAR estimated the cost of a system of this type in which the seal oil degassing vents are routed to fuel gas or compressor suction to be $22,000 (U.S. EPA, 2009). The estimated cost includes the installation of an intermediate pressure degassing drum, new piping, gas demister/filter, and a pressure regulator for the fuel line. The capital and installation costs were estimated using Guthrie's modular method of equipment cost estimation (U.S. EPA, 2009). The annual operating and maintenance cost of the systems was assumed to be minimal (U.S. EPA, 2009).

Because this option results in natural gas capture, savings can be realized from the use of the gas for beneficial purposes (e.g., the gas captured can replace other fuel that would have to be purchased). The per unit annual savings from natural gas is calculated by taking the value of the gas that is not emitted and routed to a useful purpose as a result of the capture control. This assumes that all gas that is not emitted is being routed for a useful purpose, which is reasonable given the available information on the destination of recovered seal oil degassing streams. Using the methane reductions provided in Table 4-7, the value of the natural gas saved is estimated to be $44,729 per year for centrifugal compressors equipped with one wet seal in the natural gas processing sector and $24,849 per year for centrifugal compressors equipped with one wet seal in the transmission/storage sector. These cost savings assume the value of the natural gas saved is $4/Mscf and the natural gas has a methane content of 92.8%.

Natural Gas STAR estimated the potential cost benefit of installing a seal oil capture system that uses the captured gas to fuel onsite boilers and heaters (U.S. EPA, 2009). The report estimates the potential gas savings from reduced site fuel gas consumption to be 63,000 Mscf/yr (U.S. EPA, 2009). At $4/Mscf, the potential cost savings from reduced fuel consumption would be $252,000 per year, not including the capital cost of the seal oil gas capture system.

The ICF International study (ICF, 2014) calculated emission control cost curves ($/Mcf of methane reduced) using their 2018 projected methane emission estimates (discussed in Section 3.8.2 of this document). The report evaluated the cost of preventing emissions from the use of centrifugal compressors with wet seals by capturing the seal oil degassing stream from a small disengagement vessel and recycling it back into the compressor suction (or for us as high pressure turbine fuel or low pressure fuel gas to heaters) (assuming up to 99% control of methane emissions) using their 2018 projected methane emission estimates. Their analysis assumed capital costs of $33,700 (for seal oil gas separator, seal oil gas demister for low quality gas, and seal oil gas demister for high quality gas), minimal annual operating costs, and annual product revenue benefits per centrifugal compressor of $120,000 (assuming $4/Mcf of gas) due to the reduction of product loss to the atmosphere. The estimated payback period for this control option was estimated to be three months. In total, the study estimated that methane emissions would be reduced by 19.1 Bcf (397,567 tons) methane/yr nationally. The study also estimated that VOC emissions would be reduced by 72,800 MT (or approximately 80,226 tons) nationally at a cost of $806/ton of VOC reduced.

5.0 SUMMARY

The EPA has used the data sources, analyses and studies discussed in this paper to form the Agency's understanding of vented VOC and methane emissions from centrifugal and reciprocating compressors and the applicable emissions mitigation techniques. The following are characteristics the Agency believes are important to understanding this source of VOC and methane emissions:

- Reciprocating compressors may be found throughout the oil and natural gas sector. Centrifugal compressors are predominantly used in the processing and transmission segments.

- The net 2012 methane emissions reported for compressors for the 2014 GHG Inventory were 86,259 MT for the natural gas production segment, 724,295 MT for the natural gas processing segment, and 1,261,080 MT for the natural gas transmission and storage segment, for a total of 2,071,633 MT of methane.

- Reciprocating compressor emissions may be controlled by periodic replacement of rod packing systems. Additionally, new technologies are being used that capture these emissions and route them back to the process, both reducing emissions and providing an economic benefit.

- Centrifugal compressor emissions may be controlled by using dry seals in place of wet seals. Dry seal centrifugal compressors have lower emissions, require less maintenance, and are more energy efficient that wet seal centrifugal compressors and the cost of the two technologies is similar.

- When wet seal centrifugal compressors are used, it may be feasible to capture emissions from the seal oil and route the recovered gas back to the compressor or another process, or combust the gas. Routing the gas back to a process reduces the loss to the atmosphere and reduces the destruction of natural gas.

6.0 CHARGE QUESTIONS FOR REVIEWERS

1. Please comment on the national estimates of methane emissions and methane emission factors for vented compressor emissions presented in this paper. Please comment on the activity data and the methodologies used for calculating emission factors presented in this paper.

2. Did this paper appropriately characterize the different studies and data sources that quantify vented emissions from compressors in the oil and gas sector?

3. Did this paper capture the full range of technologies available to reduce vented emissions from reciprocating compressors and wet seal centrifugal compressors at oil and gas facilities? In particular, are there other options for reducing emissions at existing reciprocating or centrifugal compressors? For example, the EPA is aware of "low emissions packing" for reciprocating compressors but has no detailed information on this technology.

4. Did this paper appropriately characterize the emissions reductions achievable from the emissions mitigation technologies discussed for reciprocating compressors and wet seal centrifugal compressors?

5. Did this paper appropriately characterize the capital and operating costs for the technologies discussed for reduction of vented emissions from reciprocating compressors and wet seal centrifugal compressors?

6. If there are emissions mitigation options for reciprocating and centrifugal compressors that were not discussed in this paper, please comment on the pros and cons of those options. Please discuss the efficacy, cost and feasibly for both new and existing compressors.

7. Are there technical limitations that make the replacement of wet seals with dry seals impractical at certain existing centrifugal compressors?

8. Are there technical reasons why an operator would use a wet seal centrifugal compressor without a gas recovery system?

9. Are there technical limitations that make the installation of gas capture systems at certain reciprocating compressors impractical?

10. Please comment on the prevalence of the different emission mitigation options in the field.

11. Given the substantial benefits of dry seal systems (e.g., lower emissions, less maintenance, and higher efficiency), are you aware of situations where new wet seal centrifugal compressors are being installed in the field? If so, are there specific applications that require wet seal compressors?

12. Are there ongoing or planned studies that will substantially improve the current understanding of vented VOC and methane emissions from reciprocating and centrifugal compressors and available techniques for increased product recovery and emissions reductions?

7.0 REFERENCES

American Petroleum Institute (API) and America's Natural Gas Alliance (ANGA). 2012. Characterizing Pivotal Sources of Methane Emissions from Natural Gas Production. Summary and Analysis of API and ANGA Survey Responses. Final Report. September 21, 2012.

Clearstone Engineering, Ltd., 2006. *Cost-Effective Directed Inspection and Maintenance Control Opportunities at Five Gas Processing Plants and Upstream Gathering Compressor Stations and Well Sites.* 2006.

EC/R, Incorporated. 2011. Memorandum to Bruce Moore from Heather Brown. *Composition of Natural Gas for Use in the Oil and Natural Gas Sector Rulemaking.* EC/R, Incorporated. June 29, 2011.

El Paso Corporation. 2010. Comments from El Paso Corporation on the Proposed Rule for Mandatory Reporting of Greenhouse Gases: Petroleum and Natural Gas Systems. *Federal Register*, Vol. 75, No. 69, Docket ID No.EPA-HQ-OAR-2009-0923.

Gas Research Institute (GRI)/U.S. Environmental Protection Agency. 1996a. National Risk Management Research Laboratory. *GRI/EPA Research and Development, Methane Emissions from the Natural Gas Industry, Volume 8: Equipment Leaks.* Prepared for the U.S. Department of Energy, Energy Information Administration. EPA-600/R-96-080h. June 1996.

Gas Research Institute (GRI)/U.S. Environmental Protection Agency. 1996b. *Research and Development, Methane Emissions from the Natural Gas Industry, Volume 8: Equipment Leaks.* June 1996. (EPA-600/R-96-080h).

REM Technology Inc. and Targa Resources. 2012. *Reducing Methane and VOC Emissions.* Presentation for the 2012 Natural Gas STAR Annual Implementation Workshop.

REM Technology Inc., et al. 2013. *Profitable Use of Vented Emission in Oil & Gas Production.* Prepared with support from the Climate Change and Emissions Management Corporation (CCEMC).

URS Corporation/University of Texas at Austin. 2011. *Natural Gas Industry Methane Emission Factor Improvement Study, Final Report.* December 2011. http://www.utexas.edu/research/ceer/GHG/files/FReports/XA_83376101_Final_Report.pdf.

U.S. Energy Information Administration (U.S. EIA). 2010. Annual U.S. Natural Gas Wellhead Price. Energy Information Administration. Natural Gas Navigator. Retrieved December 12, 2010. Available at http://www.eia.doe.gov/dnav/ng/hist/n9190us3a.htm.

U.S. Energy Information Administration (U.S. EIA). 2012a. Total Energy Annual Energy Review. Table 6.4 Natural Gas Gross Withdrawals and Natural Gas Well Productivity, Selected Years, 1960-2011. http://www.eia.gov/totalenergy/data/annual/pdf/sec6_11.pdf.

U.S. Energy Information Administration (U.S. EIA). 2012b. Total Energy Annual Energy Review. Table 5.2 Crude Oil Production and Crude Oil Well Productivity, Selected Years, 1954-2011. http://www.eia.gov/total energy/data/annual/pdf/sec5_9.pdf.

U.S. Energy Information Administration (U.S. EIA). 2013. Annual Energy Outlook 2013. Available at http://www.eia.gov/forecasts/aeo/pdf/0383%282013%29.pdf.

U.S. Environmental Protection Agency (U.S. EPA). 2006a. *Lessons Learned: Reducing Methane Emissions from Compressor Rod Packing Systems.* Natural Gas STAR. Environmental Protection Agency. 2006.

U.S. Environmental Protection Agency (U.S. EPA). 2006b. U.S. Environmental Protection Agency Lessons Learned Document. *Replacing Wet Seals with Dry Seals in Centrifugal Compressors.* October 2006. Available at http://epa.gov/gasstar/documents/ll_wetseals.pdf.

U.S. Environmental Protection Agency (U.S. EPA). 2009. *Natural Gas STAR, Partner Update, Prospective Projects Spotlight – Potential New Opportunity: Seal Oil Degassing Vent Recovery and Use, Spring 2009.* http://www.epa.gov/gasstar/documents/ngspartnerup_spring09.pdf.

U.S. Environmental Protection Agency (U.S. EPA). 2011a. *Methodology for Estimating CH4 and CO2 Emissions from Petroleum Systems. Greenhouse Gas Inventory: Emission and Sinks 1990-2009.* Washington, DC. April 2011.

U.S. Environmental Protection Agency (U.S. EPA). 2011b. *Oil and Natural Gas Sector: Standards of Performance for Crude Oil and Natural Gas Production, Transmission, and Distribution.* Background Technical Support Document for Proposed Standards, July 2011. EPA-453/R-11002.

U.S. Environmental Protection Agency (U.S. EPA). 2011c. Gas STAR PRO No. 104. *Install Automated Air/Fuel Ratio Controls.* 2011. Available at: http://epa.gov/gasstar/documents/auto-air-fuel-ratio.pdf.

U.S. Environmental Protection Agency (U.S. EPA). 2011d. Gas STAR PRO No. 203. *Pipe Glycol Dehydrator to Vapor Recovery Unit.* 2011. Available at: http://epa.gov/gasstar/documents/pipeglycoldehydratortovru.pdf.

U.S. Environmental Protection Agency (U.S. EPA). 2012. *Oil and Natural Gas Sector: Standards of Performance for Crude Oil and Natural Gas Production, Transmission, and Distribution.* Background Supplemental Technical Support Document for Proposed Standards, April 2012.

U.S. Environmental Protection Agency (U.S. EPA). 2013. Greenhouse Gas Reporting Program, Petroleum and Natural Gas Systems, Reporting Year 2012 Data. Data reported by facilities as of September 1, 2013. http://www.epa.gov/enviro/.

U.S. Environmental Protection Agency (U.S. EPA). 2014. *Inventory of Greenhouse Gas Emissions and Sinks: 1990-2012.* Washington, DC. April 2013.

(http://www.epa.gov/climatechange/Downloads/ghgemissions/US-GHG-Inventory-2014-Chapter-3-Energy.pdf).

U.S. Environmental Protection Agency (U.S. EPA). EPA Air Pollution Control Cost Manual - Sixth Edition, (EPA 452/B-02-001).

U.S. Environmental Protection Agency/ICF International. 2009. *Methane's Role in Promoting Sustainable Development in the Oil and Natural Gas Industry*. U.S. EPA, ICF International, PEMEX, EnCana Oil & Gas, Hy-Bon Engineering, Pluspetrol, Gazprom, VNIIGAZ. World Gas Conference. October 2009. Available at:
http://www.epa.gov/gasstar/documents/best_paper_award.pdf.